Cylinder

by Jennifer Boothroyd

L Lerner Publications Company · Minneapolis

I see a cylinder.

This candle is a cylinder.

This trash can is a cylinder.

These pipes are cylinders.

These hotdogs are cylinders.

These buildings are cylinders.

Do you see cylinders?